BUILD YOUR OWN
POULTRY HOUSE
AND RUN

KATIE THEAR

BUILD YOUR OWN POULTRY HOUSE AND RUN

Published by Broad Leys Publications Limited

A catalogue record for this book is available from the British Library.

ISBN: 978-1-910632-32-1

Second Edition

Unless indicated or credited otherwise, the photographs and illustrations in the book are the author's.

Graphic Design by roger@rogerb.co.uk

Permissions & Advertising
For information about reproducing articles or content from this publication, or to explore the possibility of advertising in or contributing to future editions and other titles by Broad Leys Publications, please visit www.broadleyspublications.co.uk/content

www.broadleyspublications.co.uk

The Plans

These plans include detailed instructions for the construction of:

- **a poultry house** suitable for 6 to 8 chickens.
- **a run** which will fit onto the house, although it could be used with most medium sized poultry houses.
- **added information for making a bantam house and run** suitable for 3 to 4 bantams.

The poultry house can easily be scaled down for a smaller number of birds. For a larger flock of say, a dozen birds, the house can be stretched in length. Alternatively, a second house could be built, particularly if the house needs to be moved frequently, as weight is a crucial factor.

If frequent moving is required, handles can be added at each end to facilitate this. Wheels can also be utilised; this is the best system of all but you will need to design a system that meets your requirements.

The house can also be used for ducks or a couple of geese. You won't need the pop-hole or perches for them, and you can also probably dispense with the nest boxes. The waterfowl will walk in and out through the large front door. They will appreciate a small ramp as they don't manage steps very well. Bear in mind that ducks and geese may need extra ventilation. This can be provided by ensuring that the meshed window is left open. Additional ventilation holes can also be made if necessary.

The bantam house has been designed to be made from just one sheet of plywood so it is very cheap to construct. Although detailed plans are not provided for this, it is simple to construct and incorporates the run that is described here.

The only other necessities are a feeder and a drinker. These are available in plastic or galvanised metal from poultry equipment suppliers.

It is worth mentioning that a sheet of sturdy, woven plastic can be placed on the floor of the house, with wood shavings or sawdust on top. This protects the wooden floor and makes cleaning out easy, for it is then simply a matter of removing and emptying the sheet onto the compost heap. The sheet can then be hosed down and left to dry before returning it to the house with new bedding.

Finally, every poultry house should be checked and treated regularly for red mites which can hide in cracks, coming out at night to feed on the perching birds. Perches and nest boxes can be removed for regular cleaning. Vets and many poultry suppliers sell anti-mite products.

Poultry House for 6 - 8 Hens

This is an easy to build poultry house designed to hold 6 to 8 hens. It can be adapted for 6 ducks by omitting the perch and pop-hole.

It has been constructed using 12mm exterior grade plywood. This material, proofed against the weather, will last a long time. You can economise with cheaper plywood, but if you use shuttering ply, make sure that the 'x' or ungraded side is on the inside of the house.

All timber used in the internal construction has been bought planed, therefore it will be square. The smooth surface is also less likely to harbour mites. Exterior timber for the floor bearers and roof trim is used sawn.

The house was constructed by two people in 5 hours, using power tools. Without power tools, it could take 6-7 hours to make.

The house under construction. (Thanks to Stephen Rogers)

Tools Required
Panel Saw, Tenon Saw, Jig Saw, Hand Drill (or Electric Drill), 3mm Drill Bit, 7mm Drill Bit, 19mm Hole Saw - Electric Drill (or Hand Brace and 19mm Twist Bit), Tape Measure, Square, Screwdriver, Pencil, and Sandpaper.

Tools not essential but make the job easier!
1.0m Flat Steel Rule or Metal Straight Edge, Circular Saw, Electric Sander, Battery Drill/ Screwdriver.

Materials List

Timber Requirements

3 of 8ft x 4ft (2440mm x 1220mm) sheets of 12mm exterior grade plywood.

1 of 2.4m length of 75mm x 50mm sawn treated timber.

2 of 2.7m lengths of 25mm x 38mm sawn treated timber.

4 of 3.0m lengths of 38mm x 38mm planed all round (or par) treated softwood.

2 of 2.4m lengths of 25mm x 50mm planed all round (par) treated softwood.

2 of 3.3m lengths of 12mm x 50mm planed all round (par) treated softwood.

Fixtures and Fittings

2 of 380mm
'T' hinges (zinc plated or black japanned).

3 of 50mm
butt hinges (zinc plated or black japanned).

4 of 50mm
turnbuttons (zinc plated or black japanned).

30 of M6 x 50mm
zinc plated bolts and nuts.

55 of 8 x 1.5in
countersunk zinc plated screws.

25 of 8 x 1in
countersunk zinc plated screws.

30 of 6 x 0.75in
countersunk zinc plated screws.

30 of 6 x 0.5in
countersunk zinc plated screws.

12 of 8 x 2.5in
roundhead zinc plated roofing screws (with plastic caps).

Roofing

2 of 2.0m
lengths of black *Onduline.*

Cutting List

Floor

75mm x 5mm sawn - 3 of 800mm feet.

Front

38mm x 38mm par - 2 of 1220mm top and bottom battens.
38mm x 38mm par - 2 of 795mm side battens.
25mm x 50mm par - 2 of 1020mm top and bottom ventilation window runners.
25mm x 50mm par - 2 of 580mm door battens.
12mm x 50mm par - 2 of 550mm door weatherstrip sides.
12mm x 50mm par - 1 of 620mm door weatherstrip top.
12mm x 50mm par - 2 of 900mm corner trim.

Back

38mm x 38mm par - 2 of 1220mm top and bottom battens.
38mm x 38mm par - 2 of 745mm side battens.
12mm x 50mm par - 2 of 850mm corner trim.

Pop-hole side

38mm x 38mm par - 1 of 875mm top batten (trim to fit).
38mm x 38mm par - 1 of 835mm bottom batten.
25mm x 50mm par - 2 of 760mm pop-hole runners.

Nest box side

38mm x 38mm par - 1 of 875mm top batten (trim to fit).
38mm x 38mm par - 1 of 835mm bottom batten.
12mm x 50mm par - 2 of 265mm nest box flap sides.
12mm x 50mm par - 1 of 500mm nest box flap top.

Roof

25mm x 38mm sawn - 2 of 1200mm roof batten sides (trim to fit).
25mm x 38mm sawn - 2 of 1450mm roof batten front and back (trim to fit).

Instructions

All measurements in millimetres.

3. Cutting the Sheets of Plywood

Cut three sheets of plywood as shown. Accuracy in these cuts is vital at this stage.

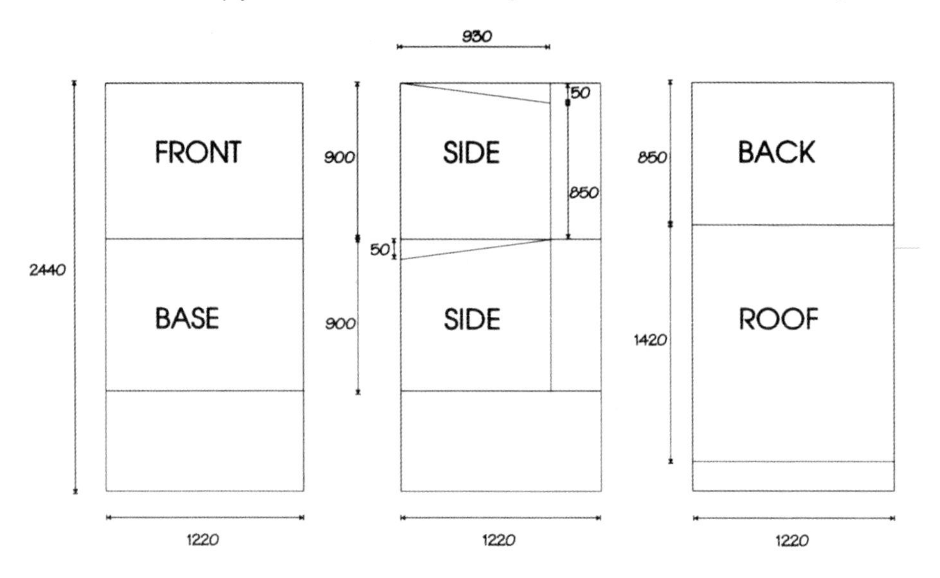

2. Making the Floor

Cut 3 pieces of 75mm x 50mm sawn timber for the floor bearers. Place on base as shown. Draw around them. Remove, drill five 3mm holes for screws for each and screw on with the 8 x 1.5in screws.

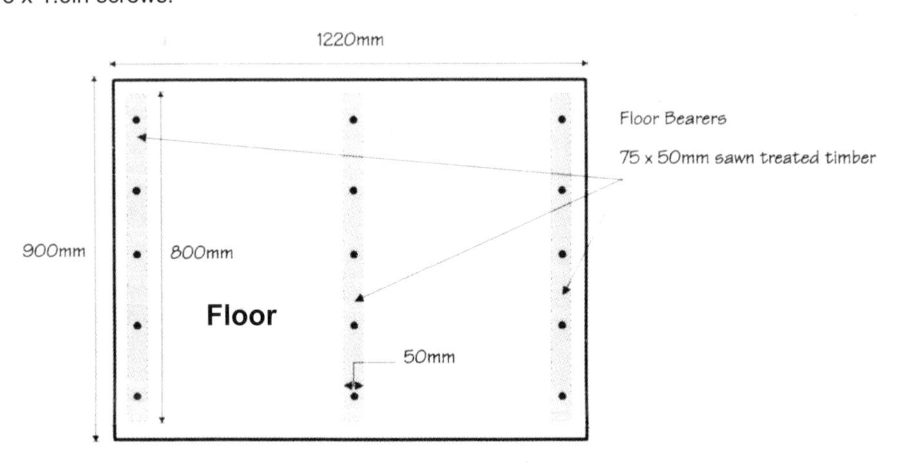

3. Making the Front

This part involves the most work. Mark out with pencil, the door and ventilator holes and the position of the screw holes into the centre line of the 38mm x 38mm timber, as shown on the plan. Drill the holes in the plywood. The curves on the door were made with the aid of a 100gm coffee jar lid.

Cut 150mm in from the bottom, up the two sides of the door.

Cut out ventilator slot and sand edges.

Screw the bottom 1220mm batten to the base, as shown, 40mm from the bottom edge. Clamping it into place will help.

Attach the top batten flush with the edge in the same way, then cut and screw on the two side battens, also flush with the edge.

Position and attach the two hinges on the outside of the door with 6 x 0.5in screws. Use just enough to hold them in place. Remove the upper hinge, and cut the remainder of the hinge side and along the top, but leave the rest of the door until later. Replace hinge using existing screw holes.

Cut out the rebate in the runner for the ventilation window. Cut two equal lengths and attach lower runner on the outside with five 8 x 1in screws from the inside. Cut sliding panel from the smallest plywood offcut 500mm x 190mm and make sure it slides in the runner. Now attach second runner leaving a 5mm gap at the top, allowing for slight expansion in damp weather. Sand edges.

Cut ventilator wire (1in x 0.5in Twilweld) to size and staple to inside of the window. Now cut out the remainder of the door and, on the inside, attach weatherstrips, using the 50mm x 12mm battens cut to fit, as shown on the plan. Position with one third of the wood as shield and two-thirds against the wood. Fix with 6 x 0.75in screws.

Cut the two 580mm battens to strengthen the door and fix them to the inside, using 8 x 1in screws. Attach two turnbuttons on the door using 6 x 0.5in screws.

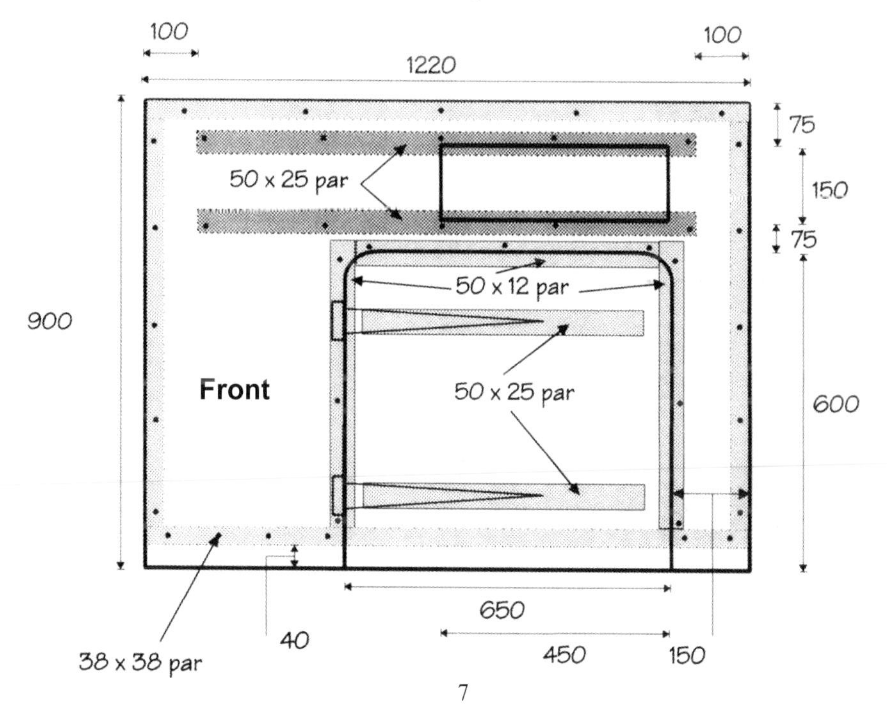

4. Making the Sides

Drill screw holes for top and bottom battens through both of the sides whilst held together.

5. Nest Box Side

Attach the bottom batten 40mm up from the bottom edge, and the top one flush with the top edge. The ends of these battens should be 50mm in from each side; you will have to cut the top batten at a slight angle to achieve this.

Draw position of flap for removing eggs from nest box and attach three of 50mm hinges, using six 0.5in screws, then cut out the nest box, removing and replacing each hinge in turn as you go. The hinges are on the outside, at the bottom.

Add weatherstrip around the inside of the nest box door, screw on a handle from the back, made from an offcut of batten, and add two external turnbuttons at the top to complete.

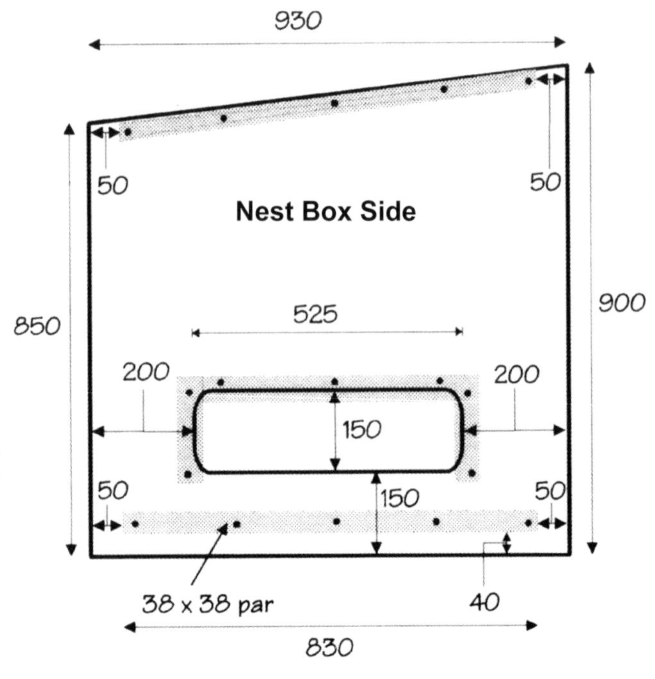

930
25 x 50 par
50
50
250
Pop-hole Side
900
850
150
350
50
50
100
40
Sliding Pop-hole on INTERNAL vertical guides

6. Pop-hole Side

Mark out pop-hole as shown. Screw on top and bottom battens, as the previous side. Cut out the pop-hole and sand edges.

The pop-hole door slides up and down on the inside of the wall, falling into place by its own weight. Thus, it does not need a handle and predators cannot 'nose' it open.

With rebated batten, attach two 760mm pop-hole runners on the inside. Insert one runner first.

Cut door from plywood offcut to 280mm x 390mm. Attach second runner, again leaving a 5mm gap at the side to allow for slight expansion.

7. Making the Back

This is a mirror image of the front, although slightly lower in height.

Draw lines 15mm from the edge for screw holes along top and sides. Attach four 38mm x 38mm battens with 1.5in screws.

Drill 19mm air holes along top as shown, 100mm down from the top edge.

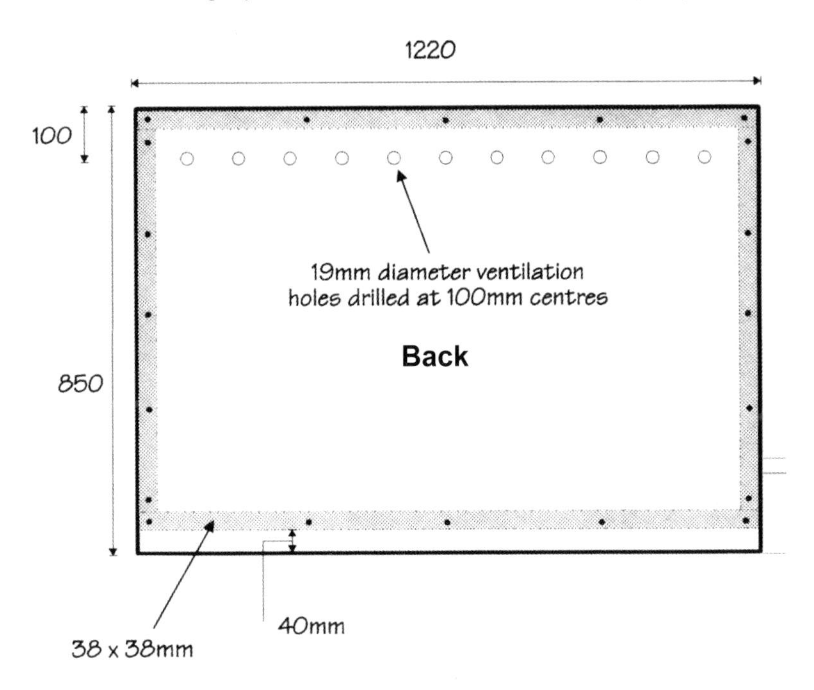

8. Assembling the House

Position the front of the house onto the base and drill four holes through the base batten of the front, through the floor.

Bolt together with nuts upwards, except the one in the
middle of the door. This has the nut and end of the bolt underneath to prevent snagging. The remaining nuts around the box should be inside the house, to protect them from weather, and for dismantling the house in future.

Attach back in exactly the same way. Sides should fit in neatly and are also bolted to the floor.

Now drill two holes in each corner, through the sides and into the side battens of the front and back walls, and bolt together, again with the nuts on the inside.

To raise and lower the pop-hole, attach a cord to the inside of the top of the pop-hole door (with staples), pass it through a cable clip or something similar, directly above on the top batten. Drill a hole at the top of the front wall, just to the side of the ventilator, passing the cord through it. Put a small loop on the end, which can hook around a nail or screw above the ventilator.

9. Roof

Cut the roof panel, allowing a 100mm overhang on the sides. The full width of the board is 1220mm; this will allow for an overhang at the back of 100mm and one at the front, of about 190mm. The front overhang provides shade and weather protection over the ventilator.

Now draw lines along the roof, 125mm in along the back and sides, and 215mm in along the front. These measurements need to be checked carefully to ensure that you will drill into the top battens of the walls of the house; adjust the measurements slightly, if necessary.

Drill two 7mm holes on each side along these centre lines and insert bolts, again with the nuts on the inside.

Screw 38mm x 25mm sawn timber battens to the lower outside edge of the roof. Give the roof a coat of wood preserver before adding the Onduline. Make sure that the edge battens are also done.

The battens around the edge provide rigidity for the roof and enable you to attach the Onduline to the edge, so that in future, you can remove the whole roof without separating the Onduline.

Lay the two 2.0m panels of Onduline onto the roof, overlapping them as necessary to fit the width and allow a 50mm overhang at the front. Attach the front with the roofing screws, through the roof and into edge batten. Allowing a 50mm overhang at the back, cut off the surplus Onduline, and screw to the back batten, as with the front. Finally, cut the front corner trims to length and tack or screw them onto the edge of the plywood and conceal the construction screws.

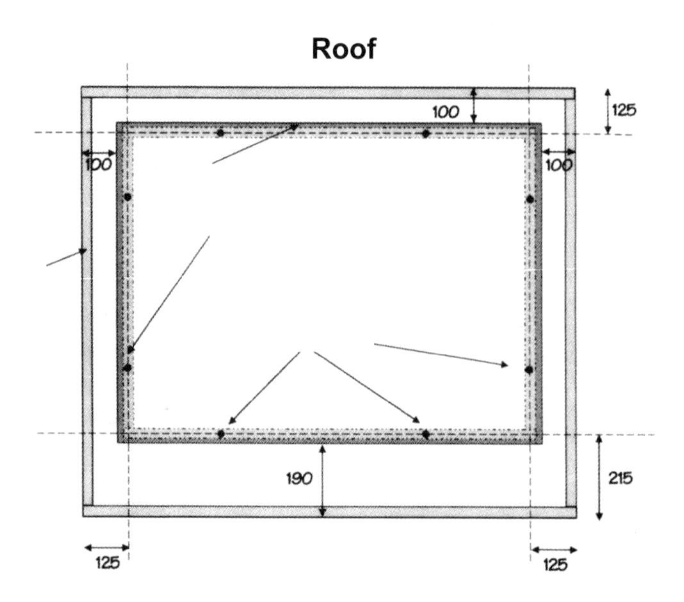

10. Nest Box

If this house is used for layers you will need a nest box arrangement - see diagram. It is not necessary to fit the internal nest box before fitting the roof; it is removeable and will fit through the door of the completed house.

Nest Box

Make from offcut plywood. Sand all edges. Glue and pin together using 1in pins (panel pins). Note: There is no back, to allow removal of eggs via nest box flap in house.

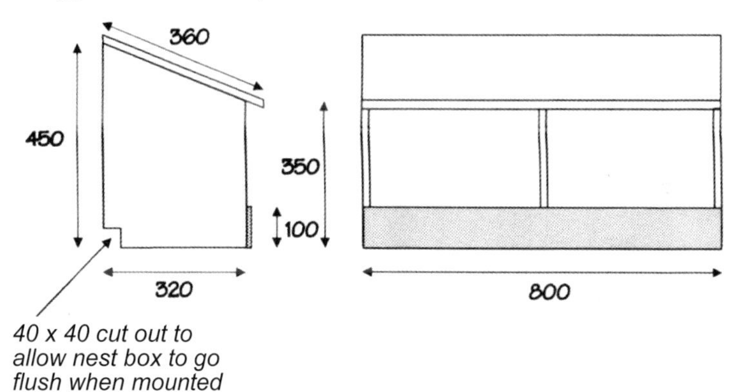

11. Perches

Perches can be free-standing, with two perches made from 38mm x 38mm sawn timber attached above a 150mm x 800mm plywood offcut. These can be removed for cleaning. The base of the perches will collect most of the droppings left in the house (see drawing).

Removeable Perches

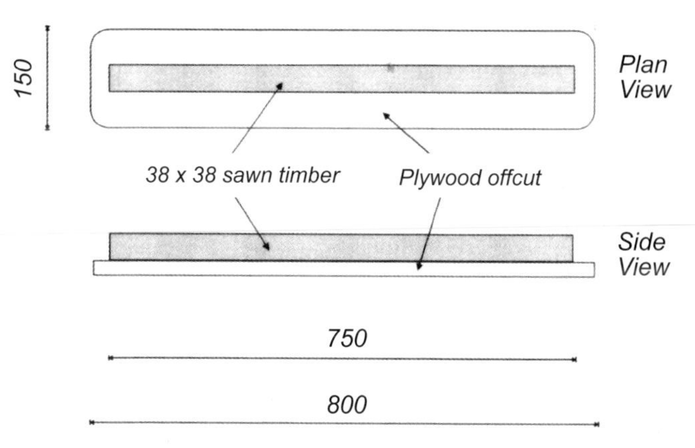

Poultry Run for 6 - 8 Hens

This run is designed to be used with the poultry house on page 4 and as pictured on the front cover. It fits flush with the side, under the roof, and is positioned so that the pop-hole opens into it. This run can also be used with other poultry houses, and is easily adapted in size to suit particular conditions.

Materials List

Timber Requirements

11 of 4.2m lengths of 50mm x 25mm sawn treated timber.

Fixtures and Fittings

2 of 50mm butt hinges (zinc plated or black japanned).
2 of 50mm turnbuttons (zinc plated or black japanned).
24 of M6 x 80mm zinc plated bolts and nuts.
84 of 8 x 3in zinc plated screws.
9m of 900mm wide 25mm Chicken wire or 25mm welded mesh (*Twilweld*).
Staples to attach the wire.

Cutting List

Side with Door

25mm x 50mm	2 of 2.4m	Top and Bottom Rails.
25mm x 50mm	4 of 850mm	Uprights.
25mm x 50mm	2 of 835mm	Door Uprights.
25mm x 50mm	2 of 485mm	Door Cross-members.
25mm x 50mm	1 of 1m	Door Brace (trim to fit).

Side without Door

As above except the parts for the door.

Ends - Parts to Make Two

25mm x 50mm	4 of 1.8m	Top and Bottom Rails.
25mm x 50mm	6 of 850mm	Uprights.

Lid

25mm x 50mm	2 of 2.4m	Side rails.
25mm x 50mm	4 of 1.8m	Cross-members (trim to fit).

Tools Required

Tenon saw, Tape Measure, Hand or Electric Drill, 3mm Drill Bit, 7mm Drill Bit, Countersink Bit, Hammer, Square, Screwdriver, Pencil, Sandpaper.

Tools not essential but make the job easier!

Mitre Saw or Mitre Box for Hand Saw, G-Clamps, Battery Drill/Screwdriver, Electric Tacker/Staple Gun.

Instructions

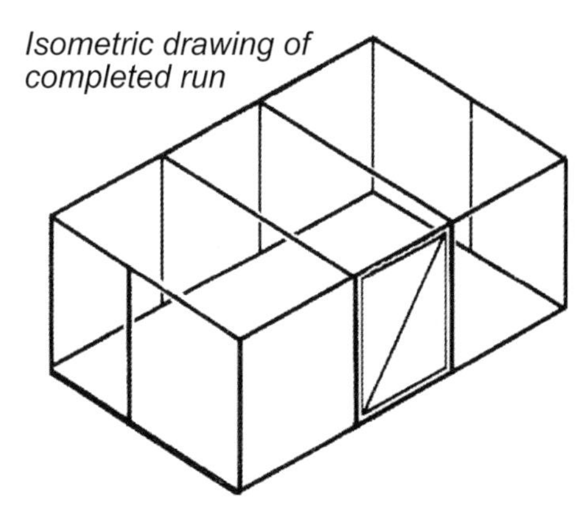

Isometric drawing of completed run

1. Carefully cut all the lengths that you require and put all of the door pieces, all of the 850mm uprights and 4 of the 1.8m lengths to one side.

2. Mark out the positions for the screw holes on all of the 2.4m and the remaining 1.8m lengths. If possible, mark all of the 2.4m length together, as this will ensure that the position of the side uprights matches the position of the lid cross-members.

3. Drill 5mm clearance holes for the 3in screws at all of the positions you have marked. Make sure that the holes are square and that they go right through the wood. Note that the two holes are always 15mm and 35mm from the end of the length of timber. Countersink these holes to a depth of about 15mm.

4. Lay out the side panel (without door) on a flat surface following the diagram. If you have a battery screwdriver, it will not be necessary to drill pilot holes for the screws in the end grain of the 850mm uprights. However, if you are assembling the panels by hand, then it will be necessary to align the pieces of wood and, using the 5mm drill, carefully mark the position of the pilot holes, which should be drilled out at 3mm.

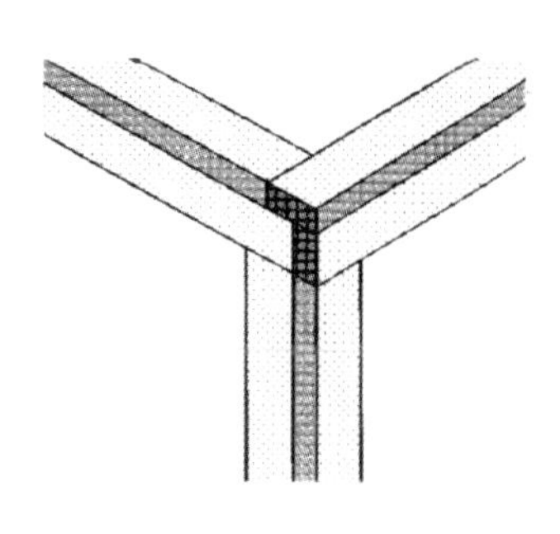

5. Ensuring that the corners are square, fix the two end uprights to the top and bottom rails. Then, fix the two central uprights in the position shown in the plan. Fixing in this sequence will tend to pull out any twist or warp in the top and bottom rails.

Poultry Run Side Without Door

2400mm

850mm

800mm

800mm

600mm

6. Make the other side and the two ends as detailed above.
Note that the upright in the pop-hole end is off-centred!

7. Make the door as if it was a small panel, ensuring that it will fit in the hole in one of the side panels that you have made. An important point to remember is that the door brace should always go up from the bottom, away from the hinges, ie, in the diagram, the hinges would be placed on the left of the door so the brace goes up and away from them. Fit the door.

8. Do not make the lid yet!

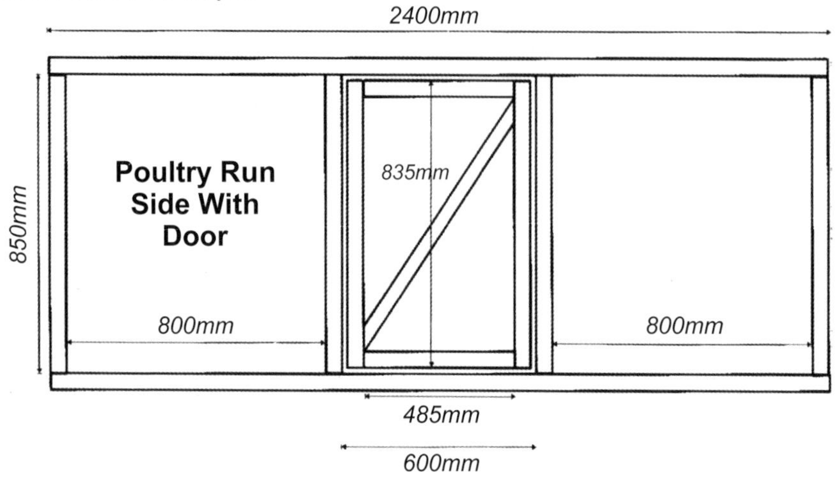

9. Using your choice of either chicken wire or welded mesh, cover what will be the inside face of the ends and sides. Make sure that if you are using chicken wire, it is pulled as taut as possible; a willing assistant is almost a necessity if you are attaching chicken wire without the benefit of an electric staple gun. Covering the panels with welded mesh is easier, although more expensive. If your poultry have to share the garden with children and footballs, then welded mesh does not deform when struck by missiles. It also remains looking as new for longer. Ensure that you do not cover the part of the end panel that will butt up to the house pop-hole.

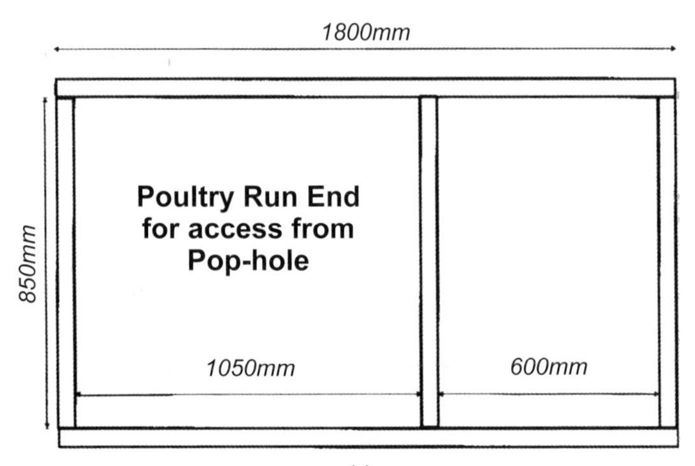

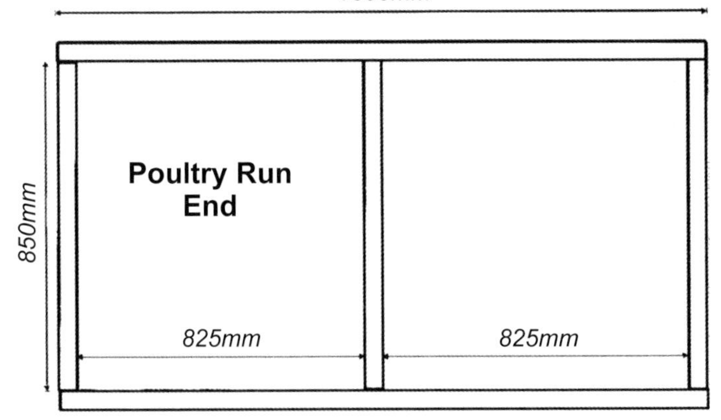

10. Join the side panels to the end panels as shown in the diagram; note that the ends go between the sides, and not vice-versa. Either clamp the corner or get an assistant to hold it together and drill two 7mm holes through the face of the side upright, into the end upright. One hole should be 250mm up from the bottom, the other 250mm down from the top. Bolt the corner together using the M6 bolts. Repeat with the other three corners. Should you want to dismantle the run for storage, it is a good idea to smear the bolts with light grease or *Vaseline* prior to tightening them.

11. The lid is constructed in-situ and has to be removed to cover it with wire. Drill and bolt one lid side to the top of one side of the run, using four equally spaced bolts. Repeat with the other side. Carefully measure the gap between the two lid sides (it should be about 1750mm) and trim the remaining four 1.8m lengths to fit. Making sure that the corners of the run are square, screw the four lid cross-members into position. Drill and bolt each lid end to the run ends using 3 equally spaced bolts per end.

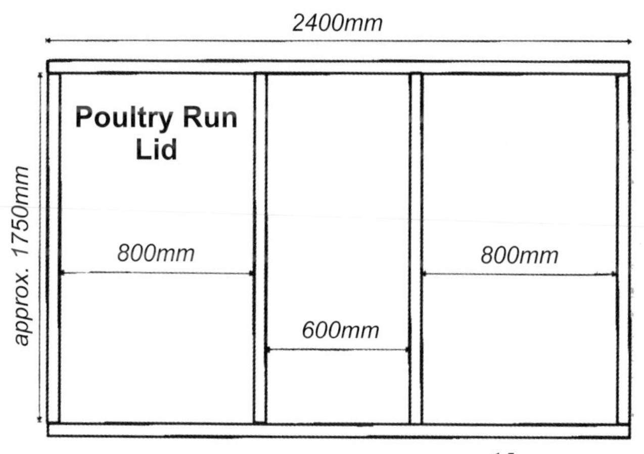

12. Remove the completed lid and cover the inside (underneath) face with wire. The lid can then be refitted.

13. Using two pieces of off-cut timber, cut two door stops and screw them to the outside of the door at the top and bottom. Fix the turnbuttons to keep the door closed.

The run is now complete.

Good luck!

Build a Bantam House suitable for 3 - 4 birds

Simple to construct, economical to build and fits with the accompanying poultry run.

This little house, can be made from a single sheet of plywood and has been designed to the maximum dimensions that a single sheet of 8ft x 4ft (2440mm x 1220mm) will allow. It has been built onto a pen framework as a combined house and run. The run part of the construction is identical to the one shown elsewhere in these plans.

The house measures 2ft x 2ft.8ins (600mm x 700mm) internally. The run measures 2ft.8ins x 8ft and is 3ft high (700mm x 2.4m x 900mm).

The photo to the right shows the sheet of plywood cut into the components of the house, and below the assembled house and run nearing completion.

The finished article. Decorations are optional! (Thanks to Mr J. R. Richardson).

Hen House Tips

The ideal position for a hen house is a sunny well drained area where there is also shade and wind protection. Hens do not like wide open spaces, for they have an instinctive fear of large birds of prey. Trees, shrubs, fences or walls provide a sense of security as well as weather protection. Place the house so that the pop-hole is on the side protected from the prevailing wind. Do not try to house too many birds in a given area, and move the house and run to new ground regularly. Outside feeders and drinkers should always be in a protected area, in the shade. Remember to close up the house after the birds have retired, to exclude predators. Nest boxes should be lined with wood shavings or sawdust that are sold specifically for poultry. There needs to be one nest box for every three birds. Eggs should be removed frequently to discourage the bad habit of egg eating, and the house itself cleaned regularly.

(From the book: *Starting with Chickens* by Katie Thear).

STARTING WITH BANTAMS

The reference guide for keeping bantams.

Written for those new to bantams, this well illustrated book provides an excellent introduction to the subject. It looks at all aspects of practical management and handling, including housing, choosing breeds and stock, breeding, showing and health.

An excellent book by David Scrivener who is an experienced bantam and large fowl poultry breeder and show judge, Chairman of The Rare Poultry Society.

List of Contents: Introduction, History of bantams, Getting started, Breeds, Feeding, Housing, Breeding, Successful hatchings, Showing, Health, Glossary of terms, References and further information, Index

It covers: Choosing a Breed, Traditional Breeds, Where to begin, What you need to know, Housing and handling, Feeding and management, Showing and breeding.

STARTING WITH DUCKS

This book covers all aspects of keeping, breeding and rearing ducks, for show, pleasure or commercial purposes. Presented in a clear concise way with an emphasis on practical detail.

Ducks are hardy and adaptable. They can be kept on any scale as long as the proper conditions are provided for them. These include access to clean water, shelter, food and good management.

List of Contents: Preface, Introduction, About the duck, Origin, classification and types, Access to water, Housing, fencing and predators, Breeds – domestic and ornamental, Illustrated description of breeds. Buying stock, Feeding, Eggs, Breeding, rearing and sexing, Table ducks, Showing, Health, Daily and seasonal care, Ducks miscellany, References and further information, Index.

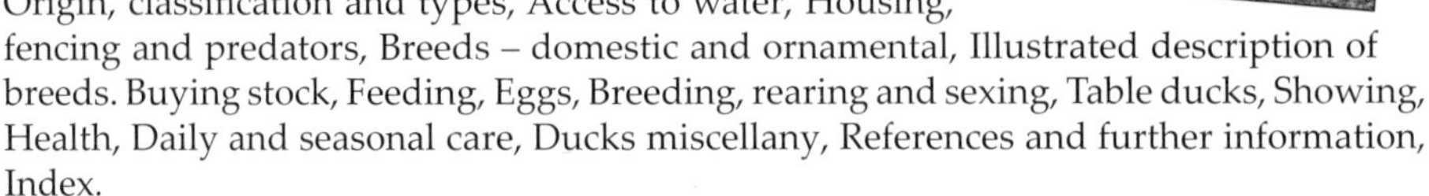

The reference guide for those keeping ducks.

www.broadleyspublications.co.uk

NOTES